Zine-Eddine KARRICHE

Field measurement and identification of solid materials

Zine-Eddine KARRICHE

Field measurement and identification of solid materials

Methods of measuring non-contact stress and strain fields and identifying solid materials

ScienciaScripts

Imprint
Any brand names and product names mentioned in this book are subject to trademark, brand or patent protection and are trademarks or registered trademarks of their respective holders. The use of brand names, product names, common names, trade names, product descriptions etc. even without a particular marking in this work is in no way to be construed to mean that such names may be regarded as unrestricted in respect of trademark and brand protection legislation and could thus be used by anyone.

Cover image: www.ingimage.com

This book is a translation from the original published under ISBN 978-620-3-41501-8.

Publisher:
Sciencia Scripts
is a trademark of
Dodo Books Indian Ocean Ltd., member of the OmniScriptum S.R.L Publishing group
str. A.Russo 15, of. 61, Chisinau-2068, Republic of Moldova Europe
Printed at: see last page
ISBN: 978-620-4-17107-4

TABLE OF CONTENTS

Introduction

The emergence of optical dimensional methods has brought new approaches and new perspectives in experimental solid mechanics. These measurement systems are very different from the classical systems such as strain gauges, extensometers and displacement sensors. The classical systems measure local information, are in direct contact with the structure but benefit from a standardization which allows them to be favoured by the industry. Systems based on optical methods, and in particular by stereo-correlation of digital images, the grid method, the image correlation method, interferometry, the infrared method and the thermographic method offer field measurements, without contact with the structure. In fact, these techniques allow access to three-dimensional displacements and deformations at any point on the surface of the structure being evaluated. These new means of measurement, due to the great wealth of information (qualitative and quantitative aspects) that they provide, renew many aspects of experimental mechanics and are favoured by researchers to complement or supplant conventional systems.
The implementation of these new field measurement techniques and their impact on experimental solid mechanics is a growing area of development, the objectives are:
Mastering the use and exploitation of field measurement techniques;
Observation of measurement fields to validate mechanical tests and models;
The use of measurement fields to identify material behaviour laws.

The presented work contributes on the one hand to the improvement of the understanding of the different optical measurement systems and on the other hand to the identification of the behaviour of materials from field measurements.

This brief is organized into 3 chapters:

Chapter I: Field measurement methods

Chapter II: calculation of 2D and 3D deformations

Chapter III: Identification from field measurements

Chapter I: Field measurement methods.

The identification of the mechanical behaviour of materials has become a major issue in many fields. This knowledge requires more and more powerful experimental means. Gauges, extensometers and other displacement sensors are powerful, easy to use and allow real-time measurements with a high degree of accuracy. Nevertheless, these techniques remain punctual and require a contact with the measurement surface which may alter the phenomenon to be observed. This is why these techniques are increasingly being replaced or supplemented by optical measurement techniques. These techniques have great advantages such as the absence of contact, a high spatial resolution and the obtaining of field information (as opposed to point information), hence the name "field measurement".

I.2 : Image correlation

The image correlation technique is a non-contact optical method for measuring 2D kinematic fields. It consists in matching two digital images of the observed plane surface in two distinct states of deformation (figure 1.1), a so-called "reference" state and a so-called "deformed" state [sutton00]. The images are spatially discretized by a CCD (Charge Coupled Device) sensor and a gray level value is obtained for each pixel. The homologous pixel of the first image is determined in the second by optimizing a correlation coefficient on their neighborhood. The point and its neighbourhood are called the correlation window.

I.2.1 : Surface coding

Image correlation is based on two major assumptions. The first assumption is that the grey levels from the initial image to the deformed image are preserved. It is therefore necessary to ensure that the mechanisms responsible for the deformation of the material do not lead to a change in contrast that would cause a risk of de-correlation. The very basis of the image correlation technique requires that the small pixel domain be identified and identifiable from other domains in the same image. Since the identification takes place through the grey levels of the pixels, the distribution of these grey levels must be sufficiently relevant in the image and at the scale of a sub-domain. The second assumption is that there must be contrasts in the image and that the distribution of these contrasts is unique in a neighborhood around the position of the homologous subdomain to be found (Figure 1.1).

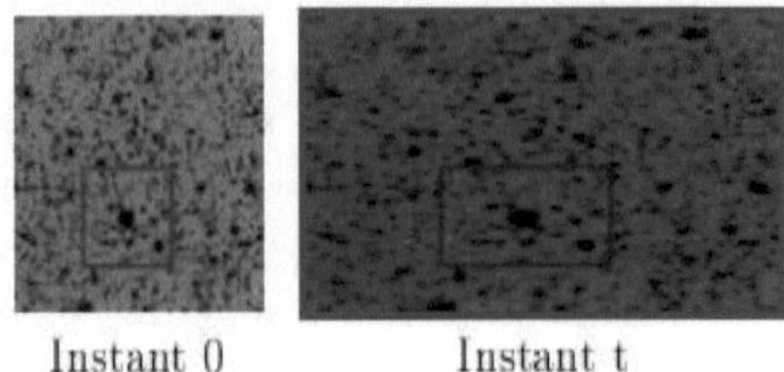

Figure 1.1: Image correlation between the initial state and the deformation state

If the surface is naturally textured (Figure 1.2 (a)), both these assumptions are directly verified. If the surface is not naturally textured, a paint is applied to the surface using an aerosol can (figure 1.2 (b)), an airbrush or other means.

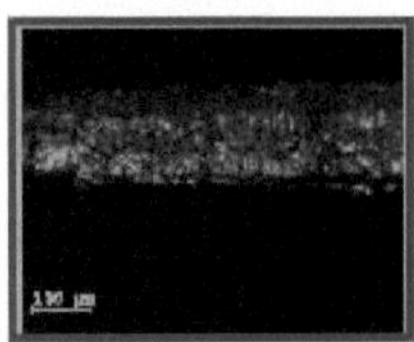

(a) (a) : natural spotting (steel) (b) : artificial spotting (spray can)

Figure (1.2): Presentation of two types of speckles

I.2.2: Form function

The reference image is partitioned into small domains D called correlation windows, over which the mechanical transformation ϕM can be approximated locally by a function ϕ called the shape function [schreier02, Bornet09]. Let us consider two signals:

_ f ($\underline{x}$) is the gray level function in the reference image

_ g ($\underline{X}$) corresponds to the gray level function in the distorted image with $\underline{x}$ = (x; y) representing the coordinates of the point of interest in the reference image and $\underline{X}$ = (X; Y) corresponding to the coordinates of the point of interest in the distorted image. Let ϕM, such that ϕM ($\underline{x}$) = $\underline{X}$be the mechanical transformation in the vicinity of any point M. The conservation of the optical flow gives :

$$g(\phi M(\underline{x})) = f(\underline{x})$$

The objective of the image correlation technique is to determine the transformation

Mechanics ϕM, knowing f and g. Thus, this problem of determining the optical flow is an ill-posed inverse problem that requires resolution by approximation [horn81].

Let $\underline{x}_0$ =(u0 , v0) a point of the reference image (usually the center of the correlation window D), $\underline{x}$ = (u, v) any point in this window and $\underline{\Delta x} = \underline{x}\underline{x}_0$ the distance of the point $\underline{x}$ to the center of the window. At this point$\underline{x}$the transformation is written :

$$\underline{X} = \phi M\,(\underline{x}) = \underline{x} + \delta\,(\underline{x}) \qquad (I.1)$$

δ ($\underline{x}$) corresponding to the displacement at point$\underline{x}$. By writing the Taylor expansion of

δ ($\underline{x}$) in the neighborhood D of the point $\underline{x}_0$, we obtain an expression ϕ of ϕM as a function of the local variations of δ at different orders :

$$\phi(\underline{x}) = \underline{x} + \delta(\underline{x}_0) + \frac{\partial\delta}{\partial x}(\underline{x}_0).\underline{\Delta x} + \frac{1}{2}\,\underline{\Delta x}^T.\frac{\partial^2\delta}{\partial X^2}(\underline{x}_0).\underline{\Delta x} + \ldots\ldots.. \qquad (I.2)$$

The approximations ϕ of ϕM obtained by dividing this expression, define the different methods encountered in image correlation:

Order 0 (2 coefficients : a1; a2) : rigid transformation. This transformation is used

In the case of small deformations. It is defined by :

ϕ($\underline{x}$) = (a1 , a2) (I.3)

Order 1: affine transformation (6 coefficients: a1 to a6) or bilinear (8 coefficients:

(a1 to a8). It is a method used in large homogeneous deformations, it is defined by :

δ($\underline{x}$) = (a1 + a2 + a3ΔX + a4ΔX + a5ΔX+ a6ΔX + a7ΔXΔY + a8ΔXΔY). (I.4)

I.2.3: Correlation coefficient

The parameters $_{ai}$ of the shape function are determined by minimizing a correlation coefficient C(ϕ).

$$_{ai} = \arg.\min C(\phi(_{ai})) \qquad (I.5)$$

This coefficient corresponds to the differences in grey level distribution that exist in the correlation window D between the initial state and the deformed state.

Several coefficients have been proposed [chambon03] as follows:

For any function a, (a) = $\int_D a(\underline{x})dx$ then $\bar{f}$ = (f) and $\bar{g}$ = (g(ϕ)) (I.6)

The most frequently encountered coefficients are written
The sum of the differences squared :

$$c_1(\phi) = (f-g(\phi))2) \quad (I.7)$$

The sum of the normalized squared differences :

$$c_2 = ((f-\bar{f}) - (g(\phi) - \bar{g})^2) / \sqrt{(f-\bar{f})^2} \cdot \sqrt{(g(\phi)-\bar{g})^2}$$

(I.8)

The centralized normalized auto-correlation function

$$c_3 = 1- (f-\bar{f}) \cdot (g(\phi) - \bar{g}) / \sqrt{(f-\bar{f})^2} \cdot \sqrt{(g(\phi)-\bar{g})^2}$$

(I.9)

All these coefficients cancel out when the two correlation windows are identical and there is no stress variation i.e. $\phi=\phi_M$. (I.10)

I.2.4 : Grey level interpolation

The integration used to compute the correlation coefficient described above is replaced by discrete sums over all pixels in the window D for any function a :

$$(a) = \sum_{i\in D} a(\underline{x}i) \qquad (I.11)$$

With $\underline{x}$i the position of the ith pixel, with integer coordinates.

Starting from the coordinates of $\phi(\underline{x}_i)$, a neighborhood gray level interpolation is necessary to determine g $(\phi(\underline{x}_i))$.The most frequently encountered interpolation methods are:

The nearest neighbor approximation.

Polynomial interpolation, including bilinear (4 pixels) and bi-cubic interpolation

(16 pixels).

B-spline interpolation.

The Fourier interpolation...

I.2.5: Optimization algorithm

The determination of the different parameters a_i of the shape function ϕ involves the minimization of the correlation coefficient (equation (1.5). This minimization requires the use of an optimization algorithm. The most commonly used algorithms are:

I.2.5.1 - Global optimization

A global optimization of all parameters a_i describing the shape function is performed by using different nonlinear optimization algorithms (gradient descent method, Newton-Raphson, Levenberg-Marquard, ...) **[sutton00].**

I.2.5.2-Partial optimization

This is an optimization to a limited number of parameters a_i, the translation components a1 and a2, leaving the other coefficients at a fixed, non-zero value. When this optimization is performed on a set of correlation windows, the higher order parameters of a given window are re-evaluated from the displacements of the centers of the neighboring windows. A new optimization of the parameters a1and a2 is then performed with the new values of the higher order parameters. This procedure is iterated until the translation components converge on all windows.

I.2.6: Conclusion

Image correlation has many advantages:
The ease of preparation of the surface of the object (projection of paint in a few minutes)
Seconds), when necessary;
The density of the information obtained. At best, each pixel of the CCD array Can be matched by correlation which provides a dense displacement field;
The step size of the virtual grid used to calculate the deformations from the displacements is chosen during the post-processing of the data and can therefore be adapted to the deformation gradients present;
The choice of post-processing parameters (window size, shape function) and image processing.

I.3 : Image stereo-correlation

The stereo image correlation technique is a non-contact kinematic field measurement method combining two-dimensional extensometry and stereovision, both performed by image correlation **[orteu97, Garcia 01a, Garcia 01c] Li 06].**

I.3.1 : Modeling of a camera and a stereovision

I.3.1.1 : Geometry of the camera

The process of image formation within a camera is carried out by passing from the world frame to the camera frame .the projection of the camera frame into the image plane from an application the transformation that leads to the coordinates of the image.

be a point P' with coordinates (x,y,z) in the world reference frame (figure I.4) the determination of the projection P' (u,v) of P on the image consists in solving the system: P'=F(P) (I.12)

This calculation needs to be done through a camera marker as shown in the following figure:

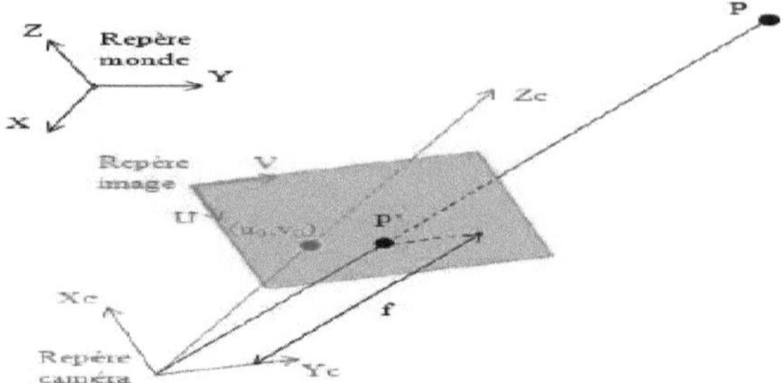

Figure I. 3 : geometry of a camera

The transformation of the coordinates of point P from the world frame to the camera frame can be considered as a combination of a rotation r and a translation t. The coordinates of P in the camera frame will be of the form :

$$\begin{bmatrix} xc \\ yc \\ zc \end{bmatrix} = T \begin{bmatrix} x \\ y \\ z \\ 1 \end{bmatrix} \qquad T= \begin{pmatrix} r11\, r12 & r13 & tx \\ r21\, r22 & r23 & ty \\ r31\, r32 & r33 & tz \end{pmatrix} = (r,t)$$

(I.13)

Such as

T: homogeneous rotation matrix and t: homogeneous translation vector

F : the focal length of the camera

I.3.2.2 : Geometry of a stereoscopic sensor

The geometric principle used for one camera remains the same with two cameras, with an additional rigid transformation linking the stereoscopic system.

Figure (I.5) shows the three rigid translations (Tg, Td, Ts) and the three reference marks.

the translations are written in the form Tg =(rg, tg) , Td = (rd , td) , Ts = (rs, ts)

where rg, rd, rs are rotations and tg, td, ts, are translations

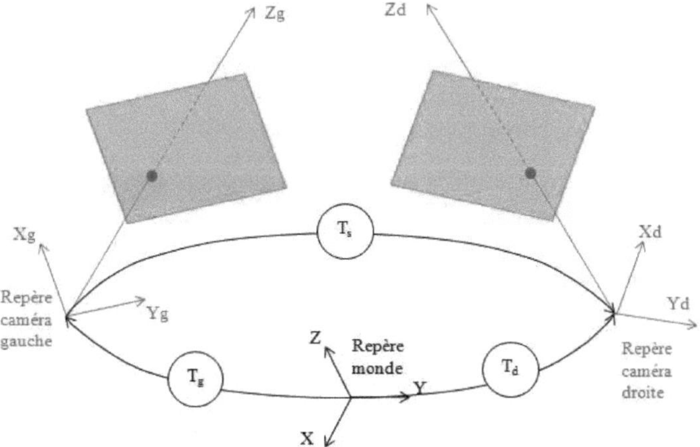

Figure I.4: three-dimensional stereovision sensor

These three rigid transformations Tg, Td, Ts, allow us to express the coordinates of a point P in the different reference frames by the following relation :

$$\begin{pmatrix} xg \\ yg \\ zg \end{pmatrix} = Tg \begin{pmatrix} x \\ y \\ z \\ 1 \end{pmatrix} \quad \begin{pmatrix} xd \\ yd \\ zd \end{pmatrix} = Td \begin{pmatrix} x \\ y \\ z \\ 1 \end{pmatrix} \quad \begin{pmatrix} xd \\ yd \\ zd \end{pmatrix} = Ts \begin{pmatrix} xg \\ yg \\ zy \\ 1 \end{pmatrix} \quad \text{(I.14)}$$

The three transformations are then linked by the following relationship:

Ts =Td T-1g (I.15)

The determination of the coordinates of the points Pd (ud, vd) and (ug, vg) in the two image planes from the coordinates of the point p(x,y,z) [Garcia 01a].

$$\begin{pmatrix} ud \\ vd \\ 1 \end{pmatrix} = \text{Kd Td} \begin{pmatrix} x \\ y \\ z \\ 1 \end{pmatrix} \quad \begin{pmatrix} ug \\ vg \\ 1 \end{pmatrix} = KgTg \begin{pmatrix} x \\ y \\ z \\ 1 \end{pmatrix} = Kg\, T^{-1}Td \begin{pmatrix} x \\ y \\ z \\ 1 \end{pmatrix}$$

(I.16)

I.3.2.Application

Because of its ease of implementation, especially in terms of surface preparation, its non-intrusiveness, and the possibility of using it at different scales, the stereo image correlation method offers a multitude of applications on different materials and structures
Composite material (carbon/epoxy specimen)

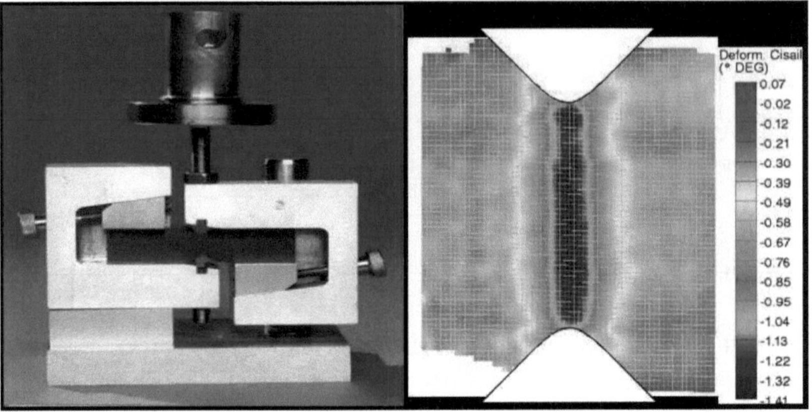

FigI. 5: shear measured by stereo-correlation on a carbon/epoxy specimen during a losipescu test **[Cazajus06].**

Field measurements by image stereo-correlation are also used on composites.

[**COMPSTON (06)**] evaluates the deformations undergone by two sandwich structures during an impact test. **[NAN04]** studies an elastomer reinforced with textile fibres ..... etc

I.4.*Speckle interferometry*

The first holographic interferometry technique that we will study is ESPI, which stands for Electronic Speckle Pattern Interferometry. It is an interferometric method using the speckle produced by the observed objects. It was first studied by Leendertz and Butters in the early 1970s. This technique takes advantage of the speckle generated by objects, so we will start by explaining its origin and its main properties.

I.4.1 Speckle phenomenon

The phenomenon of speckle is due to the microscopic relief of a surface illuminated by a monochromatic source inducing optical path variations greater than the wavelength of the light. It is produced by the interference created by the superposition of the very large number of light waves reflected at different points on this surface as shown in figure B.1. The result of this superposition of waves whose phase is randomly distributed produces a granular interference pattern called speckle. An example of a speckle pattern is shown in Figure B.2. Thus, a surface with a roughness less than the wavelength of the source will produce little or no speckle. This transition between surface

The specular or scattering effect of the wavelength used and its roughness will be studied in depth in the section.

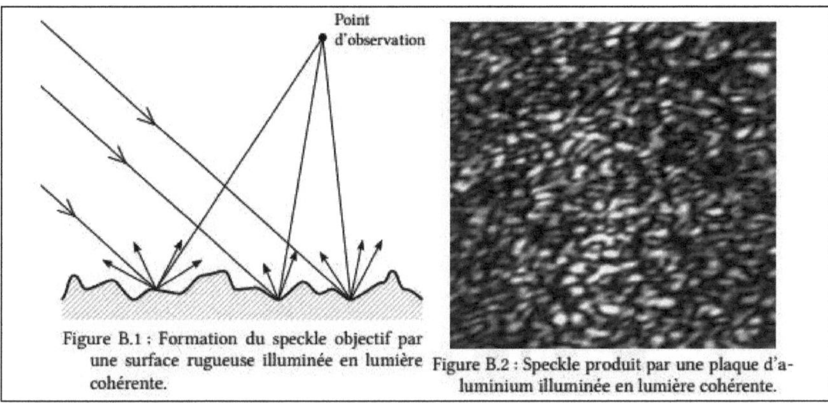

Figure B.1 : Formation du speckle objectif par une surface rugueuse illuminée en lumière cohérente.

Figure B.2 : Speckle produit par une plaque d'aluminium illuminée en lumière cohérente.

FIGI. 6.speckle formation

Despite its randomness, speckle can have statistical properties, independent of the roughness of the scatterer surface and depending only on the macroscopic parameters of the optical system used. There are two types of speckle: ***objective*** speckle and ***subjective*** speckle.

I.4.1.1 Objective Speckle

The three-dimensional average speckle grain size for a diffuser illuminated by a coherent point light source, denoted $\boldsymbol{S}$, can be estimated:

FigI.7/Three-dimensional speckle patterns

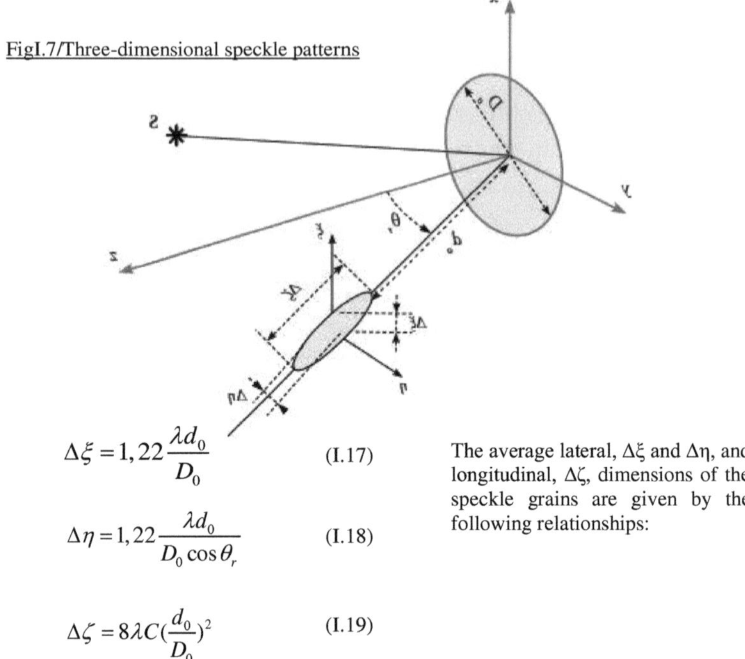

$$\Delta\xi = 1,22\frac{\lambda d_0}{D_0} \qquad (I.17)$$

$$\Delta\eta = 1,22\frac{\lambda d_0}{D_0 \cos\theta_r} \qquad (I.18)$$

$$\Delta\zeta = 8\lambda C(\frac{d_0}{D_0})^2 \qquad (I.19)$$

The average lateral, Δξ and Δη, and longitudinal, Δζ, dimensions of the speckle grains are given by the following relationships:

Where λ is the wavelength of the light, Do the diameter of the observed circular surface, *do* the observation distance, θr the observation angle with respect to the normal of the observed surface, and C a coefficient varying as a function of θr. The value of ***C*** is difficult to determine analytically for θr ≠ **0**

However, it can be calculated numerically using the curve of the evolution of the coefficient C as a function of θr :

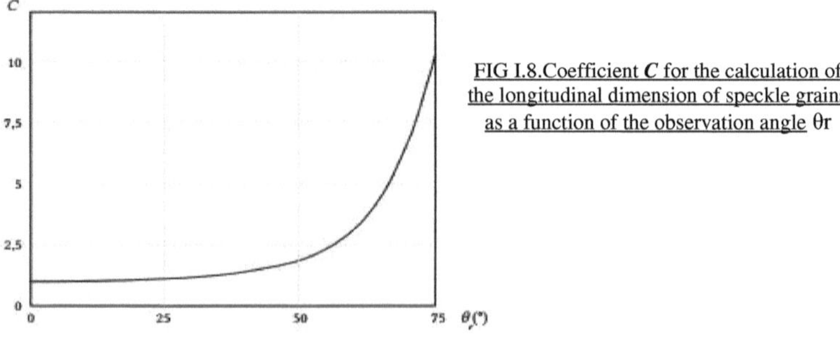

FIG I.8.Coefficient ***C*** for the calculation of the longitudinal dimension of speckle grains as a function of the observation angle θr

When the observer is in the vicinity of the axis normal to the surface, the coefficient C as well as the cosine of θr are then close to unity. The previous equations can then be simplified:

$$\Delta\xi \approx \Delta\eta \approx 1{,}22\frac{\lambda d_0}{D_0} \qquad \text{(I.19}$$

$$\Delta\zeta \approx 8\lambda(\frac{d_0}{D_0})^2 \qquad \text{(I.20)}$$

I.4.1.2 Subjective speckle

The subjective speckle, on the other hand, will depend on the optical system through which the object is observed. Indeed, the diffraction limit of the lens used must be taken into account. This implies the existence of a maximum speckle dimension for a given lens:

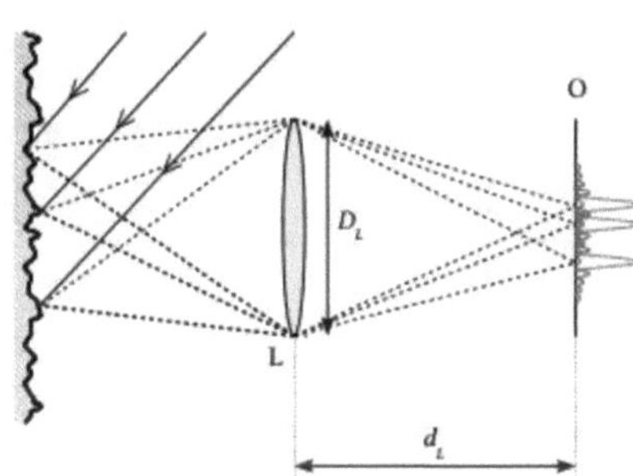

FIGI.9.Subjective speckle formation

Thus the dimensions of the speckle grains turn out to be the same as the optical resolution limit of the system, which corresponds to the dimensions of the Airy disk. The lateral dimensions of the subjective speckle grains imaged on a screen by a lens L, of diameter $_{DL}$ located at a distance ***dL*** from the screen ***O*** are given by the relation :

$$\Delta\xi \approx \Delta\eta \approx 1{,}22\frac{\lambda d_0}{D_0} \qquad \text{(I.21)}$$

NB: The d'Airy spot is the diffraction pattern resulting from light passing through a circular hole. The term Airy spot is used in the case of optical systems to qualify the best possible image of a source point by that system. A system whose impulse response gives an Airy spot is said to be diffraction limited. The name of this figure comes from George Biddell Airy, an English scientist who discovered and described the phenomenon in 1835 in "On the Diffraction of an Object-glass with Circular Aperture". The pattern is rotationally symmetric and takes the form of a bright spot surrounded by concentric circles of lower luminosity.

We can see that the calculation of the lateral dimension of the speckle grains is analogous to that of the objective speckle. Indeed, we obtain the same relation as (F.4) where we replaced the diameter of the illuminated surface by that of the imaging lens, and the distance of the surface by that of this lens. The same is true for the longitudinal size of the grains:

$$\Delta\zeta \approx 8\lambda(\frac{d_L}{D_L})^2 \qquad (I.22)$$

We also note that, in both cases, the lateral dimension of the speckle grains is directly proportional to the wavelength used. Thus, at 10 μm the grains will typically be twenty times larger than in visible light.

I.4.2 Principle of speckle interferometry

Speckle is often seen as a disturbance to be limited or suppressed. However, speckle carries information that can be used for useful purposes. Speckle interferometry exploits this information by performing interference of speckle grains, either with a reference beam or with a second illumination producing a second speckle structure interfering with the first. This interference is called a ***specklegram***. Speckle interferometry consists in numerically subtracting two specklegrams of the same object before and after displacement of its surface in order to get the measurement of this displacement. The first applications of speckle interferometry were made with photographic plates. But very quickly, in order to bypass the problems related to the long photographic development process, electronic acquisition technologies such as ***vidicon*** cameras were used. (interior view)

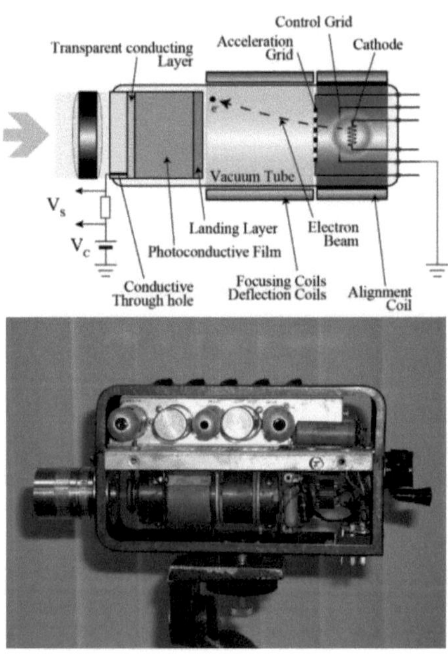

In both cases, the principle is the same. We write the complex amplitude of the object beam at any point of the sensor **:**

$$E_{obj}(x\ ,\ y)\ = o(x\ ,\ y)e^{i\phi_{obj}(x\ ,\ y)} \qquad \text{(I.23)}$$

where ***o(x, y)*** and ***ϕobj (x, y) are*** real functions representing respectively the amplitude and the phase of the complex amplitude of the object beam. The coaligned reference beam is, for its part, noted :

$$E_{ref}(x\ ,y)\ = r(x\ ,\ y)e^{i\phi_{ref}(x\ ,\ y)} \qquad \text{(I.24)}$$

where ***r(x,y)*** and ***ϕref (x,y)*** represent the amplitude and phase of the reference beam at each point of the detector. These two wavefronts are superimposed at the detector where they interfere. Since the sensor is only sensitive to the resulting intensity, the detected intensity is :

$$\begin{aligned} I_1(x\ ,\ y)\ &= |E_{obj}(x\ ,\ y)(E_{ref}(x\ ,\ y)|^2 \\ &= I_{obj}(x\ ,\ y) + I_{ref}(x\ ,y) + 2\sqrt{I_{obj}(x, y)I_{ref}(x, y)}\cos\psi(x, y) \end{aligned} \qquad \text{(I.24)}$$

I1 represents the speckle pattern, called specklegram, recorded by the sensor and **ψ(x,y)** the stochastic phase difference :

$$\psi(x\ ,\ y) = \phi_{obj}(x\ ,\ y) - \phi_{ref}(x\ ,y) \qquad \text{(I.25}$$

Expression (F.10) has been simplified using the following notations:

$$I_{obj}(x\ ,y)\ = |E_{obj}(x\ ,\ y)|^2 = o^2(x\ ,\ y) \qquad \text{(I.26)}$$

$$I_{ref}(x\ ,\ y)\ = |E_{ref}(x\ ,\ y)|^2 = r^2(x\ ,\ y) \qquad \text{(I.27)}$$

If the surface of the observed object undergoes a displacement, then the optical path is changed, resulting in a change in the phase at the sensor by a quantity denoted Δϕ ***(x, y)***. After this displacement, the object intensity is :

$$E'_{obj}(x\ ,\ y)\ = o(x\ ,y)e^{i[\phi_{obj}(x\ ,y) + \Delta\phi(x\ ,\ y)]} \qquad \text{(I.28)}$$

After superposition with the reference beam, we obtain the specklegram :

(.29)

$$I_2(x\ ,\ y)\ = I_{obj}(x\ ,\ y) + I_{ref}(x\ ,y) + 2\sqrt{I_{obj}(x\ ,\ y)I_{ref}(x\ ,\ y)}cos[\psi(x\ ,y) + \Delta\phi(x\ ,\ y)]$$

The specklegram I2 is then subtracted from the reference specklegram I1 by digital processing. The resulting intensity distribution, called the interferogram, is thus :

$$(I_1 - I_2)(x\ ,\ y)\ = 2\sqrt{I_{obj}(x\ ,\ y)I_{ref}(x\ ,\ y)}[cos\psi - cos\psi cos\Delta\phi + sin\psi\, sin\Delta\phi]_{(x\ ,\ y)} \quad \text{(I.30)}$$

$$2\sqrt{I_{obj}(x\ ,\ y)I_{ref}(x\ ,\ y)}sin[\psi(x\ ,y) + \frac{\Delta\phi(x\ ,\ y)}{2})sin(\frac{\Delta\phi(x\ ,\ y)}{2}] \quad \text{(I.31)}$$

The first factor in sine gives us the stochastic speckle noise. It is modulated by the sine of half the phase shift induced by the surface displacement. This modulation is visible if it has a much lower frequency than the high frequency speckle noise. To display these results in real time on a screen, it is necessary to use positive intensities. This is achieved by displaying the modulus | ***I1 - I2*** |.

However, this modulation does not allow us to know the exact value of the phase shift at each point. Indeed, the knowledge not of the phase shift but of the sine of its half does not make it possible to go back directly to the displacement of the surface. By comparing the result of the calculated phase with the phase shift actually produced by the object, we see that an ambiguity remains on the slope (positive or negative) of the displacement suffered by the object. In addition to this, there is the noise intrinsic to the method from the speckle represented by the first sine factor of the equation, without which we would obtain dotted curves. Without additional information on the experimental conditions of the observed displacement, it is impossible to remove this ambiguity.

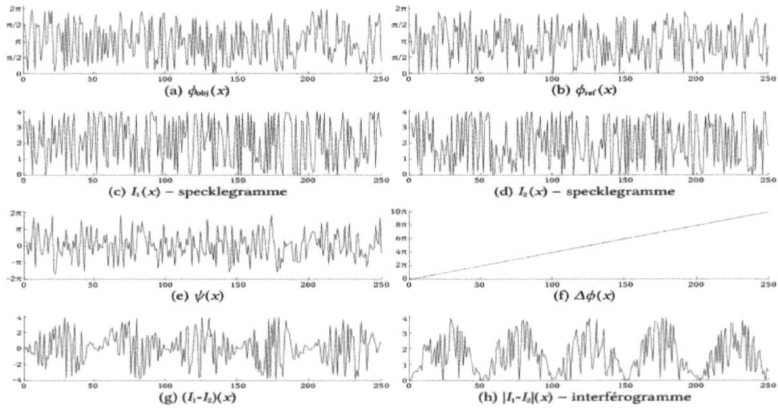

FIG.I. 11.*Numerical simulation of one-dimensional speckle interferometry*

I.4.3 *General*

Infrared thermography is a technique that allows the temperature of an observed scene to be measured at a distance and without contact.
The current trend in infrared cameras for maintenance purposes is to develop ever more sophisticated cameras:
accurate, reliable in the measurements they allow to obtain
user-friendly (especially with regard to the use of infrared images on a PC) thanks to the digitization of the image and the release of increasingly powerful software
This is partly due to new developments in cooling systems (or system temperature maintenance).
Infrared thermography has become one of the essential diagnostic tools in predictive maintenance. Indeed, most of the defects are translated by an abnormal heating or cooling. Only infrared thermography allows you to quickly observe a thermal scene and any type of object. Thus, by detecting these anomalies, often invisible to the naked eye, thermography allows corrective actions before the appearance of breakdowns or expensive problems.

The equipment

Brief description of the infrared equipment:
Non-contact infrared thermometers are designed to measure the temperature of a target at a distance. The size of the target depends on the distance between the thermometer and the target and the type of laser with which the thermometer is equipped. Depending on the degree of sophistication of the thermometer, it may or may not perform the following functions:
vary the emissivity
measure the minimum and maximum temperatures of the target
record the measured temperatures
They can be used by operators who are not specialists in infrared thermography. Thermal images, intended to visualize the temperature of an observed scene. They do not allow the measurement of temperatures.

I.4.4 Application and benefits

Thermography is commonly used indoors or outdoors for the fast and efficient control of motors, disconnectors, circuit breakers, transformers, substations and more generally of any high, medium and low voltage electrical installation. Moreover, this technique has a major advantage

In mechanics

The control of mechanical installations is an important part of predictive maintenance operations in most industrial companies. Typically, when a

mechanical part wears out prematurely, a rise in its temperature is observed, which will increase rapidly and lead to a breakdown.
In some cases, thermography becomes the only method of defect detection. For example, defects on brushes of collectors and armatures of electric motors can cause significant heating but not necessarily vibratory phenomena.

In the building

The ThermaCAM infrared camera is also a diagnostic tool for buildings. One of the most common applications is the detection of insulation problems on roofs. Water has a high thermal mass compared to roofing materials. ThermoCAM allows maintenance operators to quickly locate wet areas for effective repair.

I.4.4.1 Advantage

Thermographic conditional maintenance programs allow to locate hot spots well before they evolve into a serious situation for the company. Avoiding production stoppages, through maintenance and quality control programs, is an objective that thermography allows to reach very quickly.
Integration into predictive maintenance programs can result in significant cost savings and reduced failure rates in industrial facilities.
For electrical, mechanical, building inspection or general maintenance applications, the infrared camera provides maintenance operators with accurate thermal information to make appropriate decisions about equipment repair, plan equipment replacement and optimize facility operation.

I. 5.. Infrared thermography measurements: generalities

I.5.1 Thermal cameras

The operating principle of thermal imaging cameras is generally based on the model shown in Figure I. 12 below:

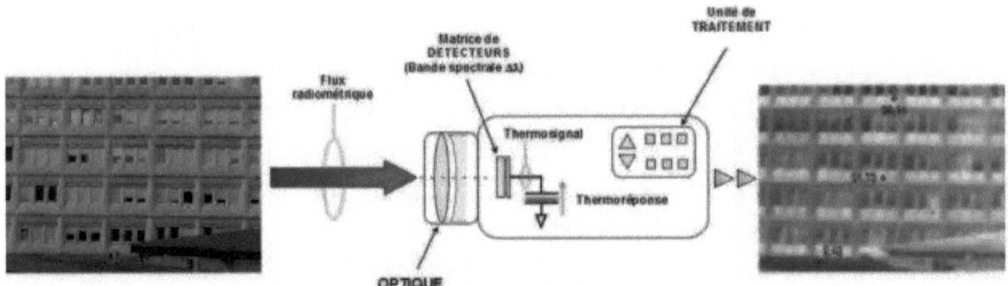

Several types of detectors are used in the form of arrays to obtain an image. Detectors are generally classified into two categories: quantum detectors and bolometric detectors.

The former are sensitive to the number of photons received and the signal they provide is proportional to this number. They can only be used when cooled. They are placed in a cryostat cooled generally by a thermal machine of the Stirling type. This type of detector can measure temperature variations of the order of a few mK (milli Kelvin). They are not very sensitive to noise.

The latter measure the heating due to the radiation received by each bolometer. A bolometer is a device that converts radiation into temperature rise. It can operate at room temperature. Their sensitivity is less. To date, they can detect temperature differences of the order of 40 mK.

The advantage of bolometric detectors is that they can be used without a cooling system. The low cost of these detectors compared to quantum detectors (as well as the absence of a cooling system) has allowed the widespread use of these cameras for thermal diagnosis.

FIG.I. 13.Temperature variation as a function of the wavelength emitted by the radiation of a building.

I.5.2 Spectral band used by thermal imaging cameras

Thermographic cameras working in the ultraviolet range will not be discussed because this spectral range offers little interest for thermography at room temperature, the object of our study. Indeed, the UV radiation is excessively weak in this temperature range. This spectral range is reserved for high temperatures where it has the advantage of a very low noise.

The spectral band of thermal cameras takes into account the transparency bands of the atmosphere. Indeed, it is imperative to be as little sensitive as possible to the thickness of the atmosphere.

I.5.3 Temperature measurement of a surface by thermography

Temperature measurement can be carried out by luminance measurement in the infrared range.

A non-contact scanning temperature measuring device is an example of a measurement using this principle. Thermal cameras use the same measurement method.

1°) Black body radiation.

At the beginning of the 20th century, Max Planck theoretically determined the emittance of an ideal body
called black body thanks to the quantum theory. The expression of this spectral emittance noted
M0 (λ, T) depends on the thermodynamic temperature T in K and the wavelength λ in m :

Plank's Law:

$$M_0(\lambda,\ T)\ =\ 2\pi hc^2 \frac{1}{\lambda^5(e^{\frac{h.c}{k\lambda T}}-1)}$$

Where h = 6.62617.10-34 J.s is Planck's constant, the speed of light in the considered medium (we will take the value of c in vacuum: c = 299792458 m.s-1) and k = 1.38066.10-23 J.K-1 the Boltzmann constant. The emittance is expressed in W.m-2 .

The spectral luminance L0 in W.m-3.sr-1 is defined as the power radiated for a given wavelength and per unit solid angle in a given direction. For a black body, this luminance does not depend on the direction considered. The black body is said to obey Lambert's law. If we integrate the luminance over a hemispherical solid angle, we obtain a relation between luminance and emittance:

$$L_0(\lambda,\ T)\ =\frac{M_0(\lambda,\ T)}{\pi}$$

The ideal black body absorbs all the incident radiation. If we integrate the total flux emitted by a black body over the entire electromagnetic spectrum and for all emerging rays, we find the total emittance ***M0*.**

As seen above, the luminance of a black body takes the form :

$$L_0(\lambda,\ T)\ =\frac{M_0(\lambda,\ T)}{\pi}=\frac{\sigma T^4}{\pi}$$

However, this equation is not applicable for thermal imaging cameras, for which only a part of the spectral band is used. For this part, it is also assumed that the surfaces

measured in the wavelength band of the camera sensitivity behave as grey bodies, so that the emissivity does not vary with wavelength and temperature (for small variations of the latter). The luminance of a surface of emissivity ε can therefore be written :

$$L(T)\ =\frac{\varepsilon\int_{\lambda_1}^{\lambda_2} M_0(\lambda,\ T)d\lambda}{\pi}$$

This calculation can be done numerically. It can be seen that the variations of the luminance of a grey body in band III and for temperatures between 300K and 360K can be written with a good approximation :

$$L(T) = \frac{\varepsilon \sigma' T^4}{\pi} \qquad (I.32)$$

Where σ' = 2.078.10-8 W.m-2.K-4 for a calculation between 8 to 14 μm.

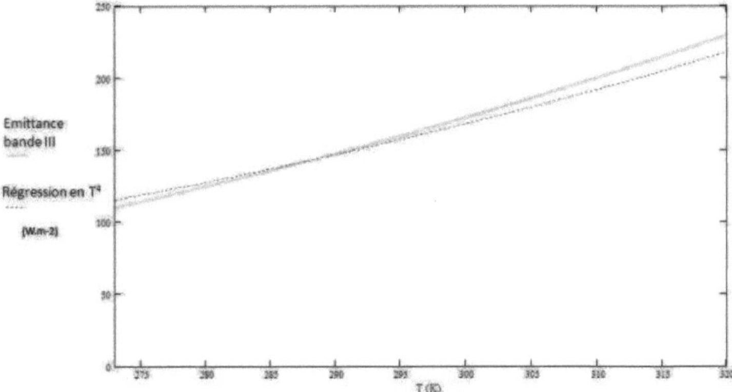

Emittance in band III in W - m-2 and its regression in T 4 as a function of temperature in K regression

Equation E.2 allows us to evaluate the influence factors of the temperature measurement using the same equations as for the total spectrum with a good approximation. The deviation on the derivative of this emittance expression is 14%. Although bolometers have a spectral response that varies little as a function of wavelength, the spectral response of the entire camera should be taken into account. This will therefore only be used in this study to evaluate the biases introduced by the other elements. The camera manufacturers integrate the properties of the cameras in their calculation code.

We will study the flux elements involved in the measurement. To determine a surface temperature, figure 1.A shows that it is necessary to know the emissivity, the atmospheric absorption and the temperature of the radiative environment. These parameters can be neglected if the aim is to carry out a qualitative diagnosis such as the identification of thermal bridges (figure 1.B)

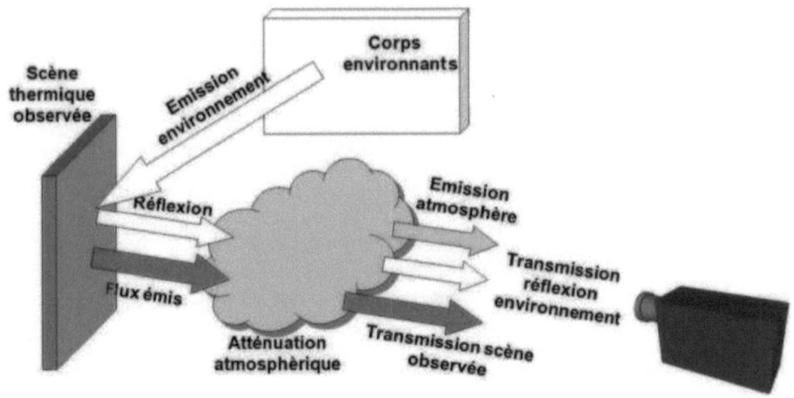

FigI. 15: Schematic of the observation of a thermal scene by IR thermography

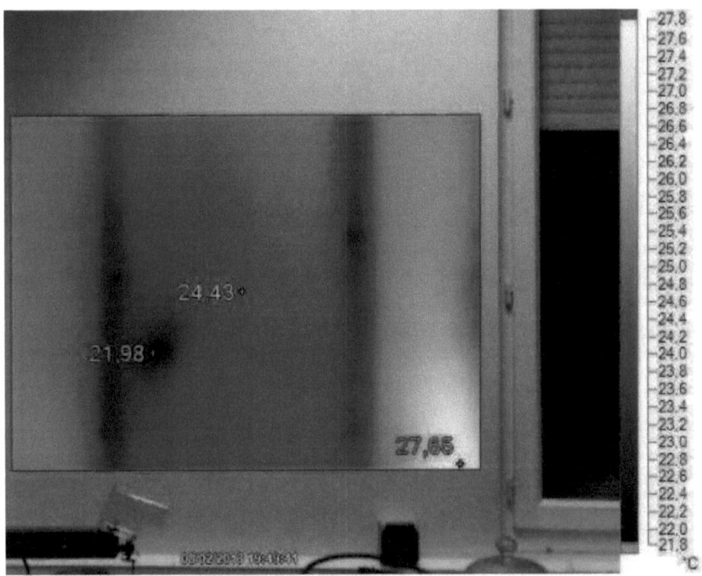

Fig. 16: Example of a thermal image showing thermal bridges due to rails of plasterboard installation

However, it is often necessary to go back to the temperature value to estimate the heat fluxes for example. For this, it is necessary to correct the images in luminance to obtain a temperature map of the observed object. From the luminance measurement, we can free ourselves from disturbing elements to determine the temperature measurement. The relationship between the luminance of the scene and the different sources of radiation allows us to estimate certain parameters. This relation is derived from the diagram in figure 1.A and is written as follows:

$$L_{mesure}(T) = \tau \cdot \varepsilon \cdot L_0(T) + \tau \cdot (1-\varepsilon) \cdot L_0(T_{env}) + (1-\tau) \cdot L_0(T_{atm})$$

where Lmesure(T) is the luminance measured by the camera, τ the transmittance of the atmosphere, ε the emissivity of the measured surface, **L0**(T) the luminance of the blackbody at temperature T, **Tenv** the temperature of the environment, and **Tatm** the temperature of the atmosphere passed through. The emissivity must be known, either by using data from the literature or by measuring it. It is difficult to determine a radiative environment temperature.

The method described in *ASTM E1862-97 is* generally preferred and suggests a solution consisting in estimating the radiation received by the wall to be studied by placing a scattering mirror in the field of the camera at the location of the wall. This mirror can be obtained by crumpling a sheet of aluminum foil and defrosting it:

FIG.I. 17.Aluminum mirror scattering according to ASTM E1862-97

However, it is necessary to ensure that the emissivity of this mirror is low enough, and that the level of ambient radiation is high enough. If this is not the case, it is necessary to know the temperature of this diffusing mirror in order to separate the clean flux of the mirror from the flux received and reflected by it.

From the hygrometry and the distance between the camera and the target object, the transmission coefficient of the atmosphere τ can be deduced. In addition, the air temperature must be known. All these parameters are fed into the surface temperature calculation code. The existing data give an atmospheric transmission which can be written as :

$$\tau = e^{-\beta \cdot x}$$

Where β is the extinction coefficient in m-1 and x is the distance in m. The value of β can be deduced from figure 1.C for an atmosphere with a hygrometry of 50%. This value depends on the spectral band; we will give for the band 3 to 5.5 μm and for the band 8 to 12 μm: $\beta_{3-5.5}$ = 3.42 - 10-3 m-1 β_{8-12} = 6.72 - 10-4 m-1.

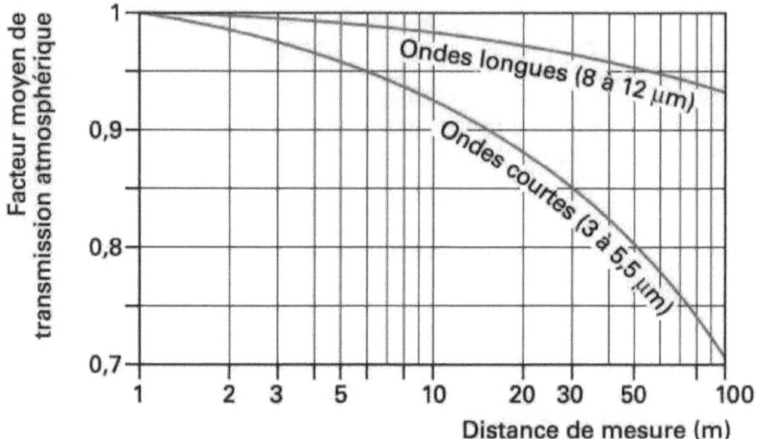

Fig.I. 18 : Typical curves of the average transmittance of the atmosphere in as a function of the measuring distance, for an atmosphere with 50% relative humidity, (according to Thermography Principles and Measurements', Engineering Technique, D. Pajani

I.5.4: Conclusion

There are now many methods of measuring displacement and deformation, and optical methods are increasingly replacing conventional point extensometer techniques. Their performance in terms of spatial resolution, ease of implementation and their absence of contact, make them robust and reliable tools. Among these techniques, image correlation and image stereo-correlation are among the most common methods, as shown by the numerous and varied applications mentioned above.

Chapter II: calculation of the deformation in 2D and 3D

The knowledge of the displacement fields is not always sufficient for the analysis of the problem and the deformation fields can appear in this case necessary. There are several methods to calculate them from the measured displacements. One of the following methods can be found.

exact point value

finite differences

This method consists in determining the value of the deformation according to the displacement values obtained on the neighbouring points and the pitch between these neighbouring points. Figure 1.3 shows the points used (in red) for the determination of the value of the green point with the finite difference method centred at 3 points for the points connected by the solid lines and at 5 points for the points connected by the dotted lines.

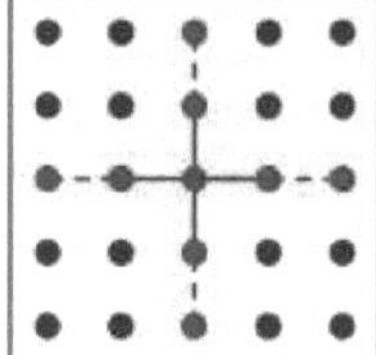

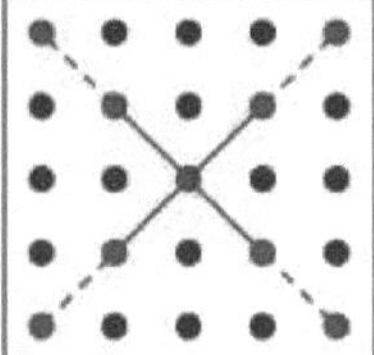

Fig II.1 : possible choice of neighbouring points

This method remains the most accurate

II.2: 3D measurement of displacement fields

The stereovision technique allows to measure the evolution of the 3D shape of an object from the recordings of several stereoscopic pairs at different states of solicitation of this object. The image correlation allows to determine the surface displacement field in the camera planes of an object from the recordings of an image at each state of deformation of this object.

The combination of these two techniques makes it possible to determine the three-dimensional displacement field at any point on the surface of the object under study.

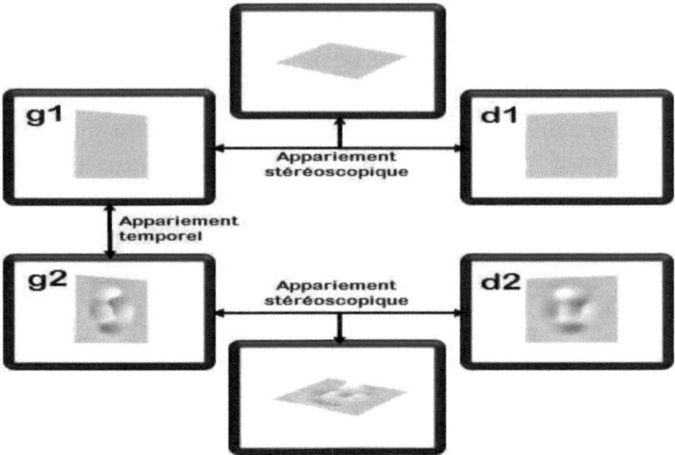

Fig (II.2): 3D displacement field determination by stereo-correlation

The principle of the method is summarized in figure (I.6) in the case of two stereo pairs (before deformation and after) in the initial state two stereo images define the object at rest by a stereo matching (step1). A second pair of images after deformation of the object, a temporal matching is performed on the images of the left camera (initial and deformed state) (step 2). Finally, a new stereoscopic matching is performed on the images of the right camera and the left camera in the deformed state (step 3)

II. 3.CONCLUSION

Both image correlation and stereo-correlation techniques allow measurements in large deformation (> 100%).

When the deformation exceeds 20%, it is difficult to perform a direct temporal matching. In this case, it is necessary to make several pairs of images between the initial state and the deformed state.

Chapter III: Identification from field measurements

II.1: Introduction

The identification of the mechanical behaviour of materials has become a major issue in many fields. This knowledge is based on increasingly powerful experimental means, using several identification techniques.

III.2: Existing identification methods

In the field of field measurements, two existing technical families of identification have been cited:

Iterative and non-iterative methods.

III.2.1 Iterative methods

III.2.1.1 **Finite element model fitting methods**

The Finite Element Modeling (FEMU) method consists of a coupling between a FE calculation code and an optimization algorithm. There are multiple variants of the original approach, these approaches are generally based on the comparison of measured and calculated forces (FEU-F approach) or between measured displacement fields and a FE calculation (FEU-U approach)

The recalibration of finite element models also has a field of application in the identification of thermomechanical models where the comparison of model and experiment involves temperature fields. This approach is called (FEU-T).

III.2.1.2 : Principle of the registration method

The finite element model fitting (FEMU) method consists of determining the parameter set (P) that reduces the difference between the quantities calculated by the finite element simulations d and the experimental measurements ^{dexp}, such that the reduction of this difference is sufficient to obtain the desired parameter set.

For this purpose, we introduce a cost function J depending on P and measuring the distance between d and ^{dexp}, i.e. : J(P) = $||d - d^{exp}||$ (II.1)

d: finite element measurements, ^{dexp}: experimental measurements

Thus the identification problem is expressed as a minimization problem, which can be solved with iterative algorithms. Figure (I.7): represents the general operation of an identification method by (REF).

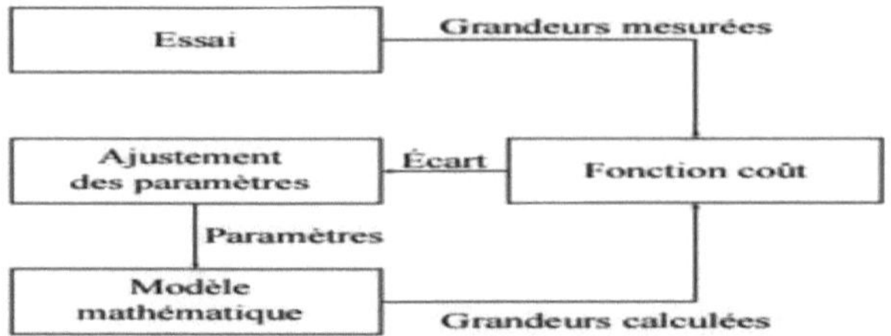

Fig (III.1): schematic diagram of the registration technique

Diagram of the principle of finite element recalibration

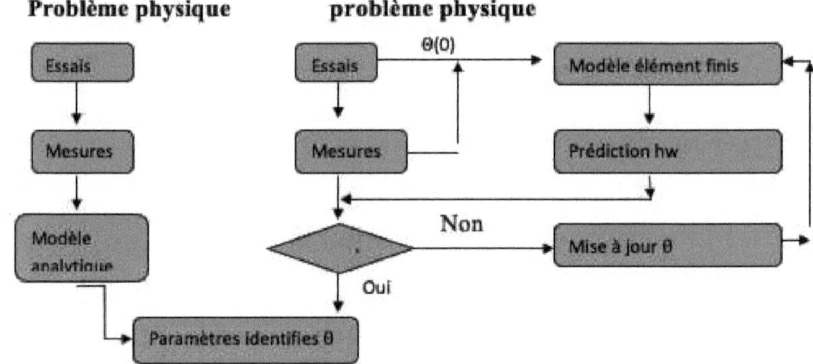

Fig (III.2) : diagram of an analytical identification and an identification by recalibration of a finite element model

m(t) : the input data (geometry, force,...)

hw(θ,t): the output variables (strain, displacement, stresses .....)

θ : is the set of parameters to be extracted (material parameter, stiffness, local force...)

θ(0): its initial estimate (a variable often needed to initialize the optimization process.

Beyond the identification strategy to be implemented, the implementation of a finite element model registration method requires two important and related points: the definition of the cost function and the choice of the method for solving the optimization problem.

The user is free to choose the output variable(s) and to associate them with a criterion to be optimized, which can be **algebraic** or **probabilistic**

III.2.1.3 : The least squares criterion

Allows the comparison of experimental data, usually tainted by measurementerrors, with a mathematical model that is supposed to describe the data.

This model can take various forms. It can be conservation laws that the measured quantities must respect. The least squares method then allows to minimize the impact of experimental errors by "adding information" in the measurement process. More precisely, finding the parameters θ minimizing the distance between m(t) **and** $h_{\varpi}(\theta,t),\ \forall t\in[0,T]$ this minimization can be written using the functional.

J such that J : P → R : that

$$\min_{\theta\in p} j(\theta)=\min_{\theta\in u}=\sum_{t=0}^{T}\frac{1}{2}j_{w}(\theta,t)^{T}\ \mathrm{v}\ j_{w}(\theta,t)\quad \text{(II.2)}$$

Where: P c ^{RP} and

Jw(θ,t) = m(t) - hw(0,t) − [j1w,j2w,........ jnw] (II.3)

With: $j_{i_{\varpi}}\ P\times[0,T]\rightarrow\square^{N}\forall i\in[0...N]$ and V the weighting matrix $N\times N$.

III.2.1.4 : The criterion of true semblance

Based on an estimate of the minimum of the variance, involves measuring an error e (t) present in equation :

m(t) = hw (θ,t) + e(t) (II.4)

And the application of a probability law .the optimal parameters minimize the following quadratic expression :

$$\min_{\theta\in P} J(\theta)=\min_{\theta\in P}=\sum_{t=0}^{T}\frac{1}{2}j_{\varpi}(\theta,t)^{T}\ \mathrm{V}\ \ j_{\varpi}(\theta,t)+\left(\theta^{(0)}-\theta\right)^{T}\mathrm{w}\ \left(\theta^{(0)}-\theta\right)\text{(II.5)}$$

These two minimization problems are usually solved iteratively:

Found θ solution of equation (II.2) and (II.5)

Such as:

$\theta^{(k+1)} = f(\theta^k)$ and $\theta = \lim_{k\to\infty} \theta^k$ (II.6)

Or the choice of **f** depends on the resolution algorithm used

III.2.1.5 : Cost function and minimization procedure

The principle of inversion resolution is to identify θ from measured data.
A general form of the cost function can be defined based on a method of calculating the difference between the experiment and the simulation. If we do not have information on the experimental errors associated with the various points of measurement, it is often assumed that they are proportional to the value of the measured quantity.
The first cost function encountered is based on the data of both the displacement U and the forces in $\hat{F}\ \Omega_t$:

Jw(θ,t) = $\hat{F}$(t) - Fint(θ,t) ∀ t ∈ [0,T] and [V= I] (II.7)

The FEMU-U (U for travel) approach.
They are also known as output methods, referring to the constraints considered as inputs to mechanical systems.
Its cost function is in the form :

Jw(θ,t) = jwu(θ,t) = $\hat{U}$(t) - Uw(θ,t) and V=Vu (II.8)

In addition, we can find cost functions based on deformations. The deformation field is obtained by a field of deformation gauges
The objective function is written as :

Jw (θ,t) = jwε(θ,t) = $\hat{\varepsilon}$(t) – ε(ϖ,θ,t) and V = Vε (II.9)

We also find a formulation in constraint :

Jw(θ,t) = jwσ(θ,t) = $\hat{\sigma}$(t) – σ(ϖ,θ,t) and V=Vσ
(II.10)

Figure (I.9) shows the difference between the simulated and experimental displacements (with the cost function of the previous equation) for a biaxial

tensile test on a composite material

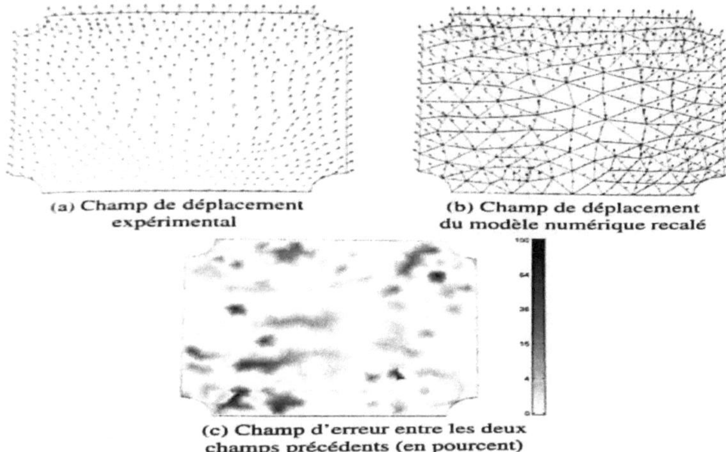

Fig (III.3): displacement fields and error after recalibration of the finite element model for a bi-axial test

III.2.6 Application in the medical field

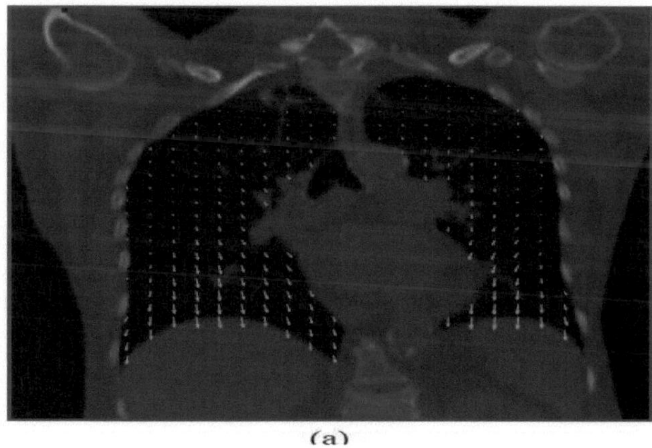

(a)

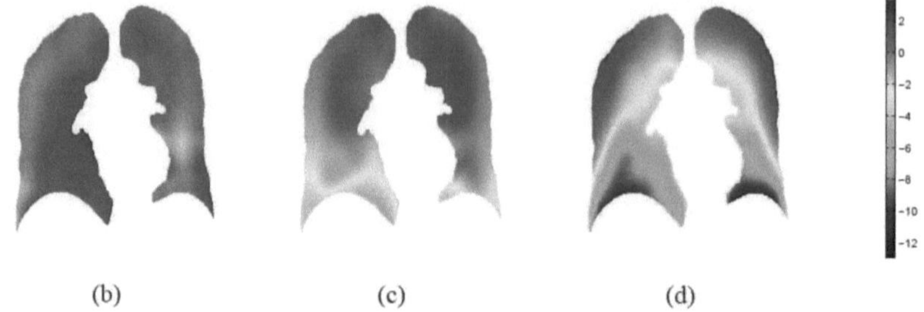

Fig. III. 4 : Displacement field obtained by shifting the image of the expiration maximum to the image of the inspiration maximum

(a) Vector visualization of the projected field on a coronal slice
(b) Transverse component ux
(c) Transverse component uy
(d) Vertical component uz

III.2.1.7: Reciprocity gap (RGM)

The reciprocity gap method consists in minimizing a cost functional constructed from the principle of virtual powers or from the reciprocity of the maxwell-betti theorem.

III.2.2 : Non-iterative method

III.2.2.1 : Virtual field method

In this section we will describe in more detail the virtual fields method which allows to identify directly the material parameters without iterative calculation. Its principle consists, for a given test, in writing the principle of the virtual works with special virtual kinematic fields well chosen. The resolution of the linear system obtained then allows access to the parameters sought. Finally, this method will be applied to the case of a particular test.

III.2.2.2: Principle

Let us consider a solid of any geometry of volume V subjected to a loading modelled by the vector $\vec{T}$ at any point of the stressed surface Sf. The principle of the work

which translate the general equilibrium of the solid is written :

$$\int_{v} \sigma : \varepsilon^{*} dv = \int_{sf} \vec{T}.\vec{u}^{*} ds \qquad \text{(III.11)}$$

Where

σ: Is the stress field and ε^*: the field of virtual deformations deduced from the kinematically admissible virtual displacements $\vec{u}^*$ kinematically admissible displacements. The principle of the virtual field method consists first of all in rewriting the above equality by introducing the behavior law. Thus, using the rules of index contraction and summation of repeated indices (x for xx, y for yy, s for xy), the stress-strain relation in the case of a plane law of anisotropic elastic type can be written :

$\sigma_i = C_{ij}\ \varepsilon_j$, i , j = x,y,s (III.12)

By transferring the previous expression into equation (II.12), the principle of virtual work will be written as follows:

$$\int_v C_{ij}\varepsilon_j\varepsilon_i^* dv = \int_{sf} \vec{T}\vec{u}^* ds \qquad \text{(III.13)}$$

If we consider that the C_{ij} are constant in the whole volume of the solid; we obtain.

$$C_{ij}\int_v \varepsilon_j\varepsilon_i^* dv = \int_{sf} \vec{T}\vec{u}^* ds \text{ (III.14)}$$

The virtual field method consists in writing equation (III.14) with as many virtual fields ($\vec{u}^*$,ε^*) kinematically admissible as the unknowns C_{ij}.

Assuming that the deformations ε are surface deformations and for well chosen virtual fields $\vec{u}^*$ the problem then comes down to determining the Cij, from a system of linear equations of the following form :

P C = R (III.15)

Where

P: a square matrix, C: a vector whose components are the Cij.

R: a vector whose components are the virtual work of the external forces per unit thickness calculated for each of these virtual fields.

The first idea is to look for virtual fields for equation (III.15) such that the final system is fully decoupled, so that the main matrix of the linear system is diagonal to the identity.

The search for a principal matrix equal to the identity matrix is necessary to have

N-1coefficients zero in equation (III.15) and only one equal to one.

N being the number of unknowns to be determined, this number N depends on the type of law that we want to identify.

Example : N=6 for anisotropic plane elasticity

N=4 for the plane orthotrope

The fields allowing to check this property are qualified as special and noted by the following $(\hat{u}^*, \hat{\varepsilon}_i^*)$. For example, if we look for the term Cpq, we write the following N equalities translating the fact that only the coefficient Cpq must appear:

$$\int_v \varepsilon_j \varepsilon_i^* dv = \begin{cases} \frac{e}{1+\delta ij} \int_{sf} (\varepsilon_i \varepsilon_j^*) + (\varepsilon_j \varepsilon_i^*) ds = 0, \forall (i,j) \neq (p,q) \\ \frac{e}{1+\delta ij} \int_{sf} (\varepsilon_i \varepsilon_j^*) + (\varepsilon_j \varepsilon_i^*) ds = 1, i = p; j = q \end{cases}$$

(III.16)

Where (e) is the thickness of the flat test piece, δij represents the Kronecker symbol. The stiffness Cij whose coefficient is unitary in the virtual work principle is then directly equal to the right-hand side in equation (II.15), i.e. the work of the forces applied with this special virtual field :

$$Cpq = \int_{sf} \vec{T}.\vec{u}^* ds \qquad \text{(III.17)}$$

When we find ourselves in more complex cases such as elasto-plasticity, no specific rule is available for the choice of the different virtual fields and an infinity of fields is then available. The idea is therefore to use N very different expressions of virtual fields chosen a priori.

III2.2.3:Application

In this section, the virtual field method is applied to a single axial tensile test. The 3mm thick specimen is made of 2024 T4 aluminium, and its geometry is described in Figure (II.1). It is assumed for the model that the upper heel is embedded and that a force of 11 000 N is applied to the lower heel in the direction given by the axis of the specimen (Figure (b)).

As explained above, let us write the principle of virtual works with particular virtual fields for a solid of any geometry of volume V subjected to a surface density of force T on its boundary Sf :

$$\int_v \sigma : \varepsilon^* dv = \int_{sf} \vec{T}.\vec{u}^* ds \qquad \text{(III.18)}$$

Where

σ: Is the stress field and ε^* the virtual strain field deduced from the virtual displacementsuu^*. The choice of the virtual fields is an essential step. It is therefore advisable to select kinematically admissible fields that allow simplifying the calculations. Thus, in our case, we can consider a first contraction field defined by the equations :

$$\begin{Bmatrix} u^{1*}_x = 0 \\ u^{1*}_y = -y \end{Bmatrix} \longrightarrow \begin{cases} \varepsilon^{1*}_x = 0 \\ \varepsilon^{1*}_y = -1 \end{cases} \qquad \text{(III.19)}$$

$$\varepsilon^{1*}_{s} = 0$$

Where

x,y and(s) are the conventional contracted indices.

Using these virtual fields, the first member of equation (III.18) becomes :

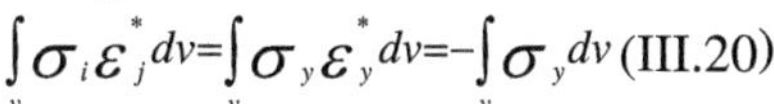

$$\int_{v} \sigma_i \varepsilon^{*}_{j} dv = \int_{v} \sigma_y \varepsilon^{*}_{y} dv = -\int_{v} \sigma_y dv \quad \text{(III.20)}$$

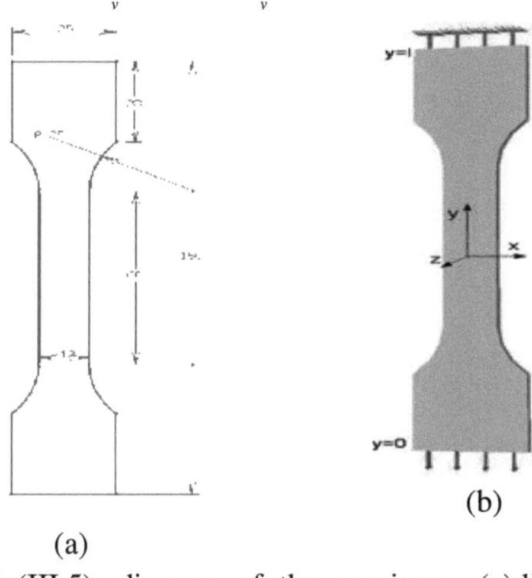

(a) (b)

Fig(III.5): diagram of the specimen, (a)dimension in mm,(b) boundary condition adopted for the problem

Given the planar nature of the system under study, the isotropic elastic behaviour law is written in the following form:

$$\begin{pmatrix} \sigma_x \\ \sigma_y \\ \sigma_s \end{pmatrix} = \begin{bmatrix} Cxx & Cxy & 0 \\ Cxy & Cxx & 0 \\ 0 & 0 & \frac{Cxx-Cxy}{2} \end{bmatrix} \cdot \begin{pmatrix} \varepsilon x \\ \varepsilon y \\ \varepsilon s \end{pmatrix}$$

(III.21)

By introducing in equation (II.20) the behaviour law described by equation (III.21),

we get

$$-\int_{v} \sigma_y dv = -\int_{v} (C_{xx} \varepsilon_y + C_{xy} \varepsilon_x) dv \qquad \text{(III.22)}$$

In the case of a homogeneous material, considering that the Cij are constant in the whole volume of the solid, we obtain :

$$-\int_{v} (C_{xx} \varepsilon_y + C_{xy} \varepsilon_x) dv = -eC_x \int_{s} \varepsilon_y dxdy - eC_{xy} \int_{s} \varepsilon_x dxdy \quad \text{(II.23)}$$

Where

e: is the thickness of the flat test piece

Assuming that the measured deformation is uniform over a material area i of surface (si), we can write

$$\left.\begin{aligned}\int_s \varepsilon_y dxdy = \sum_{i=1}^{N} \varepsilon_y^i s^i \\ \int_s \varepsilon_x dxdy = \sum_{i=1}^{N} \varepsilon_x^i s^i\end{aligned}\right\} \quad \text{(III.24)}$$

Or

N : is the number of pixels, each pixel covering the same area $^{if} = s = s_T/N$

s_T: the total measurement area

We get:

$$\left.\begin{aligned}\sum_{i=1}^{N} \varepsilon_y^i s^i = \frac{S_T}{N} \sum_{i=1}^{N} \varepsilon_y^i = S_T . \bar{\varepsilon}_y \\ \sum_{i=1}^{N} \varepsilon_x^i s^i = \frac{S_T}{N} \sum_{i=1}^{N} \varepsilon_x^i = S_T \, \bar{\varepsilon}_x\end{aligned}\right\} \quad \text{(III.25)}$$

With

$\bar{\varepsilon}_y = \frac{1}{N} \sum_{i=1}^{N} \varepsilon_y^i$ the average of the deformations according to y

Finally, the first member of equation (II. 18) is written

$$\int_v \sigma_{ij} \varepsilon_{ij}^* dv = -e S_T [C_{xx} \overline{\varepsilon}_y + C_{xy} \overline{\varepsilon}_x]$$

(III.26)

For the simple tensile test described in figure 1, the virtual work of the external forces can be written as

$$\int_{sf} T_i u_i^* ds = FL$$

(III.27)

Where F is the applied force and L is the length of the specimen.

Equation (II.18) will be written with the first virtual field equation (II.19)

$$\text{Cxx}\bar{\varepsilon}_y + Cxy \, \bar{\varepsilon}_x = -\frac{FL}{eS_T}$$

(III.28)

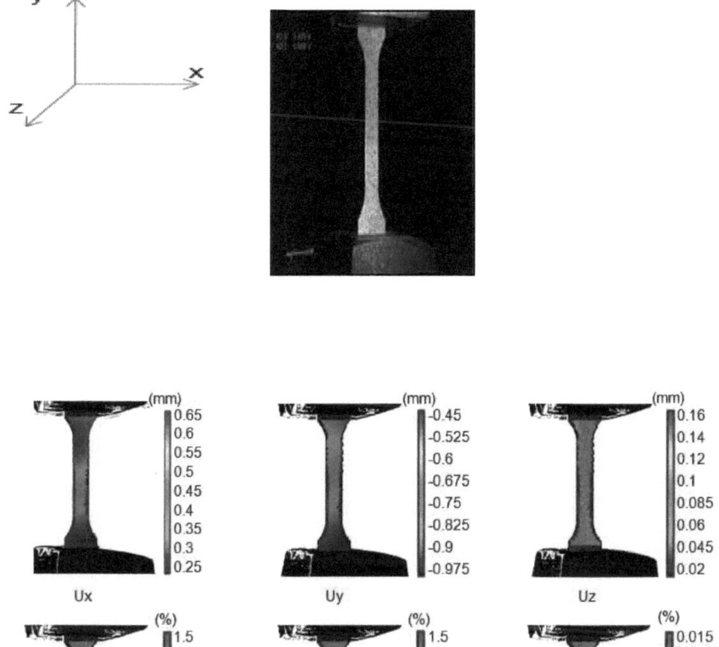

Fig III. 6: Simple tensile test in y direction on a 2024 T4 aluminium specimen

Following the same approach, the use of a second virtual field (II.29) leads to a second equation (II.30), thus forming with equation (II.28) a system of two equations with two unknowns, Cxx and Cxy (equation (II.31). This second virtual field is not strictly kinematically admissible, but the contribution of the embedding forces to the external virtual work is considered to be zero or negligible due to the symmetry of these forces with respect to the vertical axis.

$$\begin{cases} u_x^{2*} = x \\ u_y^{2*} = 0 \end{cases} \longrightarrow \begin{cases} \varepsilon_x^{2*} = 1 \\ \varepsilon_y^{2*} = 0 \\ \varepsilon_s^{2*} = 0 \end{cases} \qquad \text{(III.29)}$$

$$\text{Cxx}\ \bar{\varepsilon}_x + \text{Cxy}\ \bar{\varepsilon}_y = 0$$
(III.30)

$$\begin{bmatrix} \bar{\varepsilon}_x & \bar{\varepsilon}_y \\ \bar{\varepsilon}_y & \bar{\varepsilon}_x \end{bmatrix} \cdot \begin{pmatrix} Cxx \\ Cxy \end{pmatrix} = \begin{pmatrix} 0 \\ \frac{FL}{eS_T} \end{pmatrix}$$
(III.31)

The parameters then calculated allow us to determine the Young's modulus E and the Poisson's ratio ν of our material by the following relations:

$$\text{E= Cxx (1- ν2)} = \frac{FL}{eS_{T\bar{\varepsilon}_Y}}$$
(III.32)

$$\nu = \frac{Cxy}{Cxx} = -\frac{\bar{\varepsilon}_x}{\bar{\varepsilon}_y}$$

Using the parameters of the specimen shown in Figure 2, we obtain a Modulus of elasticity of 76447 MPa and a Poisson's ratio of 0.35. It is possible to compare the results obtained by the virtual field method with those obtained from the standard experimental characterization implemented on this type of test. The Young's modulus thus measured has the value 77232MPa, the Poisson's ratio is taken from the bibliography (table 1). We note that the two methods give close values. Due to the lack of experimental results and cases treated, we cannot make a complete comparison of this method. We can nevertheless think that its use on other types of tests is possible.

	Standard characterization	Virtual field method
Young's modulus (MPa)	77232	76447
Poisson's ratio	0.33	0.35

Tab III. 1: Identification results of the elastic parameters of 2024T4 aluminium with the virtual field method

III.2.2.4.CONCLUSION

The evaluation of kinematic field measurement techniques has favoured the development of numerous identification techniques by inverse methods, among which the virtual field method and the finite element resetting method seem to be the most suitable for the identification of parameters driving behavioural laws. The former proves to be very efficient in terms of calculation time, particularly non-iterative, but it is still difficult to apply to non-linear problems.

III.2.2.5 Summary

There are now many methods of measuring displacement and deformation, and optical methods are increasingly replacing conventional point extensometry techniques. Their performance in terms of spatial resolution, ease of implementation and their absence of contact, make them robust and reliable tools. Among these techniques, image correlation and stereo image correlation are among the most common methods, as evidenced by the numerous and varied applications mentioned above.

The identification of the mechanical behaviour of materials has become a major issue in many fields. This knowledge requires more and more powerful experimental means using several techniques of identification among which we can quote the most known.

The non-iterative method (virtual field method) and iterative method (finite element resetting method), seem to be the most suitable for the identification of parameters driving behavior laws, the first one proves to be very efficient in terms of computational field especially non-iterative it remains nevertheless difficult to apply for non-linear problems, for the second finite element resetting method remains when it, a robust and easily generalizable technique.

General conclusion

This dissertation presented the research work contributing to a better understanding of measurement systems based on optical dimensional methods such as digital image stereo-correlation, interferometry, image correlation, infrared thermography and the grid method, for a great mastery in terms of use and also in terms of exploitation of results.

In the first chapter, we gave an overview of the different techniques for measuring kinematic fields. We chose to focus on correlation and stereo-correlation of digital images, two very common techniques.

The second chapter represents the accuracy of the system for displacement measurement and deformation calculation in 2D and 3D with the finite difference method and the stereo-correlation method.

Finally, the third chapter is dedicated to the exploitation of the results of the field measurements, parametric identification techniques have been developed in order to take advantage of the largest amount of information available. Among these, the virtual field method and the finite element method are the most widespread and the most adapted to the identification of parameters driving a behavioural law. The first one, based on the principle of virtual works, uses specific and judiciously chosen virtual fields and allows to identify parameters in a direct way without iteration.

The second technique used is the finite element resetting method, which is the most commonly used, although it is costly in terms of calculation time and can be easily generalized.

In order to illustrate these two methods an application is presented in the case of an identification of the elastic parameters for the virtual field method.

The work presented in this thesis paves the way for future developments.

Concerning the characterization and evaluation of the measurement systems, we can envisage many avenues of research. Indeed, the parametric study carried out in image correlation and image stereo-correlation could be extended to other parameters such as noise, objectives, in image stereo-correlation the means available did not allow us to carry out a study from synthetic images except

The particularities of the Aramis 3D software.

Concerning the identification part from field measurements, following the first approach, there are still many tracks to explore, for the method of recalibration of finite element models it would be interesting to evaluate the performance of the algorithm and to increase its capacity to fully process the experimental data. Other avenues concern the examination of other types of behaviour to be identified, such as anisotropic elasticity.

The last important point relates to the ability to identify any three-dimensional behaviour of a material or structure from any three-dimensional test.

Bibliography

[Andrieu97] S , Andrieux, A ,Ben Abda and D on the identification of plane cracks via

The concept of deviation from reciprocity in plasticity. Proceedings of

The Academy of Sciences Series 1, Mathematics 1997

[cazajus06] V , Cazajus, S miston and M, Karama losipescu shear test on carbon/epoxy

Composites, 12th European conference on composite material , ECCM 12

Vol, 01, 46, b32, 6330, 2006

[chalal05] H, Chalal numerical and experimental identification of non

Linearization of composite materials from kinematic field measurements

D. thesis, Ecole National Supérieure d'Arts et Métiers Chalon en

Champagne, 2005

[compston06] P. compston , M styles and S , kalyanasundaram low energy impact damage

Modes in aluminum foam and polymer foam sandwich structures journal of

Sandwich structures and materials, vol, 8 pages 365-379, 2006

[cooreman08] steven cooreman. Identification of the plastic material behavior thought full

Field displacement measurements end inverse methods these de doctorat, vrije

Universities brussel, 2008

[Grédiac89] M, Grédiac , principle of the virtual traveaux and identification

.

The Academy of Sciences, series 2, vol 309 pages 1,1989

[Grédiac 90] M Grédiac and A, vautrin, A new method for determination of bending rigidities

Of thin anisotropic plates, journal of applied mechanics vol 57, pages 964-968,

1990

[Grédiac96a] M Grédiac , on the direct determination of invariant parameters governing

Anisotropic plate bending problems ,international journal of solids and

Structures , vol 33(27) pages 3969-3982-1996

[Grédiac96b] M Grédiac and P, A Paris direct identification of elastic constants of anosotropic

Plates by modal analysis theoretical and numerical aspect , journal of sound and

Vibration vol,195(3) pages 401-415 1996

[Guo07] Baoquiao guo simultaneous identification of the stiffness parameters and

Damping of isotropic plates under vibration by the method of

Virtual fields, doctoral thesis, Ecole nationale supérieure d'arts

Et métiers chalon - en - champagne ,2007

[karama06] M , karama and B lorrain , numerical and experimental modelling of the

Behaviour of sandwich structures, mechanics and industries vol 7 pages 39-48

2006

[kavangh71] KT kavangh and R, W clough finite element application in the charactersation of

Elastic solids, international journal of solids and structures, vol 7 pages 11-23

1971

[lecompte07] David lecmopte , arwen Suits Hugo sol John van tomme and Danny van

Hemelrijck mixed numerical experimental technique for orthotropy parameter

Identification using biaxial tensile tests on crucirfrom specimens international of

Solids and structures vol,44 pages 1643-1656,2007

[mulle09] M, Mulle , R, zitoune , F collombet L, Robert et y-h grunevald embedded FBG

And 3D dic for the stress analysis of a structural specimen subjected to bending of

Composite structures, vol 91 no 1 pages 48-55,2009

[nam04] T H Nam mechanical properties of the composite material with elastomeric matrix

Reinforced by textile cords , these de doctorate technical university of Liberec

Czech republic,2004

[owlabi06] gbadebo M, owolabi and mecra N,K singh, A comparison betwen two analytical

Models that approximate notch-root elastical strain-stress components in two

Phase particle reinforced, metal matrix composites under multiaxial cyclic

Loading experiments international journal of fatigue vol, 28 no,8 pages 918-925

Anjust 2006

[pirron00] Fabrice Pierron, serge zhavoronok and Michel Grédiac identification of throught

Thickness properties of thick laminated tubes using the virtual fields methods

International journal of solids and structures vol 37 no 32 pages 4437-4453

Augist 2000

[Molimard05] J, Molimard , R, le riche , A, vautrin and J,R,Lec , identification of the four

Orthotropic plate stiffness using a single open hole tensile test experimental

Mechanics, vol 4555, pages 404-411,2005

[Geers96] M G,D Greers T,R de borsts and W ,A,Mbrekelmans , cemuting strain fields from

Discrete displacement filds in 2d solids , international journal of solids and

Structure vol 33(29) pages 4293-4307 , 1996

MIX
Papier aus verantwortungsvollen Quellen
Paper from responsible sources
FSC® C105338

Printed by Books on Demand GmbH, Norderstedt / Germany